MATH
WORD PROBLEM

ADDITION

1. 3 red peaches and 19 green peaches are in the basket. How many peaches are in the basket?

2. 18 oranges were in the basket. More oranges were added to the basket. Now there are 19 oranges. How many oranges were added to the basket?

3. Marcie has 12 more bananas than Audrey. Audrey has 6 bananas. How many bananas does Marcie have?

4. Paul has 7 plums and Jake has 2 plums. How many plums do Paul and Jake have together?

5. 24 pears were in the basket. 4 are red and the rest are green. How many pears are green?

6. 4 marbles were in the basket. More marbles were added to the basket. Now there are 21 marbles. How many marbles were added to the basket?

7. Marcie has 12 more avocados than Jackie. Jackie has 8 avocados. How many avocados does Marcie have?

1. 2 red marbles and 5 green marbles are in the basket. How many marbles are in the basket?

2. 28 apples were in the basket. 17 are red and the rest are green. How many apples are green?

3. 16 apricots were in the basket. More apricots were added to the basket. Now there are 32 apricots. How many apricots were added to the basket?

4. Marcie has 8 more peaches than Jackie. Jackie has 16 peaches. How many peaches does Marcie have?

5. Adam has 6 avocados and Brian has 20 avocados. How many avocados do Adam and Brian have together?

6. 6 pears were in the basket. More pears were added to the basket. Now there are 26 pears. How many pears were added to the basket?

7. Jake has 7 balls and Billy has 5 balls. How many balls do Jake and Billy have together?

1. 4 red apricots and 14 green apricots are in the basket. How many apricots are in the basket?

....

2. 17 apples were in the basket. 7 are red and the rest are green. How many apples are green?

....

3. Jennifer has 13 more avocados than Marin. Marin has 12 avocados. How many avocados does Jennifer have?

....

4. 6 peaches were in the basket. More peaches were added to the basket. Now there are 11 peaches. How many peaches were added to the basket?

....

5. Allan has 5 oranges and Billy has 11 oranges. How many oranges do Allan and Billy have together?

....

6. Audrey has 20 more marbles than Marcie. Marcie has 5 marbles. How many marbles does Audrey have?

....

7. 8 red plums and 15 green plums are in the basket. How many plums are in the basket?

....

1. 17 red pears and 15 green pears are in the basket. How many pears are in the basket?

   ....

2. Steven has 2 avocados and Jake has 10 avocados. How many avocados do Steven and Jake have together?

   ....

3. Ellen has 17 more marbles than Jackie. Jackie has 15 marbles. How many marbles does Ellen have?

   ....

4. 10 plums were in the basket. More plums were added to the basket. Now there are 16 plums. How many plums were added to the basket?

   ....

5. 10 bananas were in the basket. 9 are red and the rest are green. How many bananas are green?

   ....

6. 5 balls were in the basket. More balls were added to the basket. Now there are 7 balls. How many balls were added to the basket?

   ....

7. 19 red oranges and 20 green oranges are in the basket. How many oranges are in the basket?

   ....

1. 12 peaches were in the basket. 1 is red and the rest are green. How many peaches are green?

2. 12 red avocados and 3 green avocados are in the basket. How many avocados are in the basket?

3. Marin has 15 more marbles than Jackie. Jackie has 10 marbles. How many marbles does Marin have?

4. 12 apricots were in the basket. More apricots were added to the basket. Now there are 31 apricots. How many apricots were added to the basket?

5. Donald has 15 oranges and Billy has 16 oranges. How many oranges do Donald and Billy have together?

6. Brian has 18 pears and Donald has 2 pears. How many pears do Brian and Donald have together?

7. 2 bananas were in the basket. More bananas were added to the basket. Now there are 18 bananas. How many bananas were added to the basket?

1. 8 red bananas and 9 green bananas are in the basket. How many bananas are in the basket?

---

2. 21 balls were in the basket. 17 are red and the rest are green. How many balls are green?

---

3. Brian has 4 apples and Paul has 16 apples. How many apples do Brian and Paul have together?

---

4. 6 pears were in the basket. More pears were added to the basket. Now there are 8 pears. How many pears were added to the basket?

---

5. Jennifer has 2 more avocados than Janet. Janet has 1 avocado. How many avocados does Jennifer have?

---

6. 7 red plums and 16 green plums are in the basket. How many plums are in the basket?

---

7. 19 oranges were in the basket. 11 are red and the rest are green. How many oranges are green?

---

1. 17 pears were in the basket. More pears were added to the basket. Now there are 23 pears. How many pears were added to the basket?

....

2. Paul has 1 ball and Allan has 20 balls. How many balls do Paul and Allan have together?

....

3. Jackie has 18 more bananas than Jennifer. Jennifer has 10 bananas. How many bananas does Jackie have?

....

4. 10 avocados were in the basket. 7 are red and the rest are green. How many avocados are green?

....

5. 5 red peaches and 5 green peaches are in the basket. How many peaches are in the basket?

....

6. 15 red apricots and 11 green apricots are in the basket. How many apricots are in the basket?

....

7. 18 oranges were in the basket. 6 are red and the rest are green. How many oranges are green?

....

1. 16 pears were in the basket. More pears were added to the basket. Now there are 24 pears. How many pears were added to the basket?

   ....

2. 6 red plums and 12 green plums are in the basket. How many plums are in the basket?

   ....

3. Allan has 5 peaches and Donald has 7 peaches. How many peaches do Allan and Donald have together?

   ....

4. Amy has 2 more apples than Audrey. Audrey has 8 apples. How many apples does Amy have?

   ....

5. 23 avocados were in the basket. 8 are red and the rest are green. How many avocados are green?

   ....

6. Billy has 11 marbles and David has 1 marble. How many marbles do Billy and David have together?

   ....

7. 7 balls were in the basket. 5 are red and the rest are green. How many balls are green?

   ....

1. Donald has 4 bananas and Steven has 13 bananas. How many bananas do Donald and Steven have together?

2. 3 red peaches and 3 green peaches are in the basket. How many peaches are in the basket?

3. 8 oranges were in the basket. More oranges were added to the basket. Now there are 20 oranges. How many oranges were added to the basket?

4. Jennifer has 10 more avocados than Marin. Marin has 14 avocados. How many avocados does Jennifer have?

5. 17 plums were in the basket. 11 are red and the rest are green. How many plums are green?

6. 17 red balls and 12 green balls are in the basket. How many balls are in the basket?

7. Marcie has 15 more pears than Jackie. Jackie has 4 pears. How many pears does Marcie have?

1. 38 peaches were in the basket. 19 are red and the rest are green. How many peaches are green?

    ....

2. 19 red plums and 18 green plums are in the basket. How many plums are in the basket?

    ....

3. Jackie has 14 more apples than Janet. Janet has 16 apples. How many apples does Jackie have?

    ....

4. Donald has 10 bananas and Allan has 8 bananas. How many bananas do Donald and Allan have together?

    ....

5. 1 ball was in the basket. More balls were added to the basket. Now there are 10 balls. How many balls were added to the basket?

    ....

6. Steven has 19 pears and Allan has 6 pears. How many pears do Steven and Allan have together?

    ....

7. 10 red apricots and 4 green apricots are in the basket. How many apricots are in the basket?

    ....

SUBTRACTION

1. 8 apricots are in the basket. 4 apricots are taken out of the basket. How many apricots are in the basket now?

....

2. Some oranges were in the basket. 14 oranges were taken from the basket. Now there are 0 oranges. How many oranges were in the basket before some of the oranges were taken?

....

3. Donald has 5 peaches. David has 6 peaches. How many more peaches does David have than Donald?

....

4. 19 bananas were in the basket. Some of the bananas were removed from the basket. Now there are 7 bananas. How many bananas were removed from the basket?

....

5. Audrey has 7 fewer balls than Michele. Michele has 18 balls. How many balls does Audrey have?

....

6. 10 marbles are in the basket. 9 are red and the rest are green. How many marbles are green?

....

7. David has 2 pears. Allan has 19 pears. How many more pears does Allan have than David?

....

1. Donald has 13 bananas. Billy has 19 bananas. How many more bananas does Billy have than Donald?

2. Some oranges were in the basket. 2 oranges were taken from the basket. Now there are 5 oranges. How many oranges were in the basket before some of the oranges were taken?

3. 14 pears are in the basket. 4 pears are taken out of the basket. How many pears are in the basket now?

4. 6 peaches are in the basket. 3 are red and the rest are green. How many peaches are green?

5. Marcie has 2 fewer marbles than Jennifer. Jennifer has 6 marbles. How many marbles does Marcie have?

6. 7 apples were in the basket. Some of the apples were removed from the basket. Now there are 6 apples. How many apples were removed from the basket?

7. Some plums were in the basket. 6 plums were taken from the basket. Now there are 0 plums. How many plums were in the basket before some of the plums were taken?

1. 13 apricots were in the basket. Some of the apricots were removed from the basket. Now there are 8 apricots. How many apricots were removed from the basket?

....

2. Amy has 2 fewer oranges than Michele. Michele has 10 oranges. How many oranges does Amy have?

....

3. Steven has 3 balls. Adam has 12 balls. How many more balls does Adam have than Steven?

....

4. Some marbles were in the basket. 8 marbles were taken from the basket. Now there are 7 marbles. How many marbles were in the basket before some of the marbles were taken?

....

5. 7 avocados are in the basket. 5 avocados are taken out of the basket. How many avocados are in the basket now?

....

6. 15 pears are in the basket. 8 are red and the rest are green. How many pears are green?

....

7. Michele has 0 fewer apples than Marin. Marin has 2 apples. How many apples does Michele have?

....

1. David has 12 apples. Donald has 13 apples. How many more apples does Donald have than David?

   ....

2. 6 marbles are in the basket. 1 is red and the rest are green. How many marbles are green?

   ....

3. 4 balls are in the basket. 1 ball is taken out of the basket. How many balls are in the basket now?

   ....

4. Some avocados were in the basket. 2 avocados were taken from the basket. Now there are 4 avocados. How many avocados were in the basket before some of the avocados were taken?

   ....

5. 1 banana was in the basket. Some of the bananas were removed from the basket. Now there are 0 bananas. How many bananas were removed from the basket?

   ....

6. Michele has 3 fewer peaches than Marcie. Marcie has 18 peaches. How many peaches does Michele have?

   ....

7. 20 plums were in the basket. Some of the plums were removed from the basket. Now there are 19 plums. How many plums were removed from the basket?

   ....

1. 9 oranges were in the basket. Some of the oranges were removed from the basket. Now there is 1 orange. How many oranges were removed from the basket?

2. Some bananas were in the basket. 3 bananas were taken from the basket. Now there are 3 bananas. How many bananas were in the basket before some of the bananas were taken?

3. Audrey has 4 fewer avocados than Sandra. Sandra has 16 avocados. How many avocados does Audrey have?

4. 18 plums are in the basket. 15 plums are taken out of the basket. How many plums are in the basket now?

5. Brian has 4 marbles. Allan has 20 marbles. How many more marbles does Allan have than Brian?

6. 14 apples are in the basket. 14 are red and the rest are green. How many apples are green?

7. 17 apricots are in the basket. 2 are red and the rest are green. How many apricots are green?

1. Some pears were in the basket. 5 pears were taken from the basket. Now there is 1 pear. How many pears were in the basket before some of the pears were taken?

2. 8 apricots were in the basket. Some of the apricots were removed from the basket. Now there are 3 apricots. How many apricots were removed from the basket?

3. 20 bananas are in the basket. 19 are red and the rest are green. How many bananas are green?

4. Marcie has 0 fewer marbles than Marin. Marin has 13 marbles. How many marbles does Marcie have?

5. Billy has 5 avocados. Steven has 15 avocados. How many more avocados does Steven have than Billy?

6. 20 balls are in the basket. 12 balls are taken out of the basket. How many balls are in the basket now?

7. Paul has 8 peaches. Donald has 8 peaches. How many more peaches does Donald have than Paul?

1. 2 pears were in the basket. Some of the pears were removed from the basket. Now there is 1 pear. How many pears were removed from the basket?

2. Amy has 11 fewer marbles than Sharon. Sharon has 12 marbles. How many marbles does Amy have?

3. Some peaches were in the basket. 6 peaches were taken from the basket. Now there are 5 peaches. How many peaches were in the basket before some of the peaches were taken?

4. 15 oranges are in the basket. 12 oranges are taken out of the basket. How many oranges are in the basket now?

5. 19 balls are in the basket. 11 are red and the rest are green. How many balls are green?

6. Steven has 2 avocados. Allan has 5 avocados. How many more avocados does Allan have than Steven?

7. 9 bananas are in the basket. 6 bananas are taken out of the basket. How many bananas are in the basket now?

1. 17 apples were in the basket. Some of the apples were removed from the basket. Now there are 2 apples. How many apples were removed from the basket?

2. 14 bananas are in the basket. 13 are red and the rest are green. How many bananas are green?

3. Some apricots were in the basket. 10 apricots were taken from the basket. Now there are 4 apricots. How many apricots were in the basket before some of the apricots were taken?

4. 4 peaches are in the basket. 4 peaches are taken out of the basket. How many peaches are in the basket now?

5. Billy has 18 plums. David has 20 plums. How many more plums does David have than Billy?

6. Audrey has 1 fewer marble than Jennifer. Jennifer has 12 marbles. How many marbles does Audrey have?

7. 17 balls are in the basket. 16 are red and the rest are green. How many balls are green?

1. Some apples were in the basket. 3 apples were taken from the basket. Now there is 1 apple. How many apples were in the basket before some of the apples were taken?

....

2. 4 oranges were in the basket. Some of the oranges were removed from the basket. Now there are 3 oranges. How many oranges were removed from the basket?

....

3. 19 pears are in the basket. 14 are red and the rest are green. How many pears are green?

....

4. 4 marbles are in the basket. 4 marbles are taken out of the basket. How many marbles are in the basket now?

....

5. Paul has 1 plum. Steven has 18 plums. How many more plums does Steven have than Paul?

....

6. Audrey has 2 fewer avocados than Sharon. Sharon has 11 avocados. How many avocados does Audrey have?

....

7. Billy has 8 bananas. David has 10 bananas. How many more bananas does David have than Billy?

....

1. 14 plums are in the basket. 1 is red and the rest are green. How many plums are green?

2. 11 pears were in the basket. Some of the pears were removed from the basket. Now there are 0 pears. How many pears were removed from the basket?

3. 20 balls are in the basket. 2 balls are taken out of the basket. How many balls are in the basket now?

4. Jackie has 6 fewer apples than Jennifer. Jennifer has 9 apples. How many apples does Jackie have?

5. Brian has 9 oranges. Jake has 17 oranges. How many more oranges does Jake have than Brian?

6. Some peaches were in the basket. 3 peaches were taken from the basket. Now there are 6 peaches. How many peaches were in the basket before some of the peaches were taken?

7. 2 bananas were in the basket. Some of the bananas were removed from the basket. Now there are 0 bananas. How many bananas were removed from the basket?

MULTIPLICATION

1. If there are two oranges in each box and there are four boxes, how many oranges are there in total?

    ....

2. Sharon's garden has six rows of pumpkins. Each row has eight pumpkins. How many pumpkins does Sharon have in all?

    ....

3. Steven swims three laps every day. How many laps will Steven swim in three days?

    ....

4. Allan can cycle eight miles per hour. How far can Allan cycle in five hours?

    ....

5. Janet has 10 times more plums than Amy. Amy has seven plums. How many plums does Janet have?

    ....

6. Jennifer's garden has nine rows of pumpkins. Each row has three pumpkins. How many pumpkins does Jennifer have in all?

    ....

7. Steven can cycle 11 miles per hour. How far can Steven cycle in seven hours?

    ....

1. Paul swims 12 laps every day. How many laps will Paul swim in two days?

....

2. Jennifer's garden has nine rows of pumpkins. Each row has six pumpkins. How many pumpkins does Jennifer have in all?

....

3. Marcie has 10 times more peaches than Ellen. Ellen has three peaches. How many peaches does Marcie have?

....

4. If there are six balls in each box and there are three boxes, how many balls are there in total?

....

5. Jake can cycle eight miles per hour. How far can Jake cycle in eight hours?

....

6. If there are three oranges in each box and there are six boxes, how many oranges are there in total?

....

7. Michele swims eight laps every day. How many laps will Michele swim in nine days?

....

1. If there are seven oranges in each box and there are three boxes, how many oranges are there in total?

....

2. Amy's garden has 11 rows of pumpkins. Each row has four pumpkins. How many pumpkins does Amy have in all?

....

3. Allan has two times more apricots than Janet. Janet has five apricots. How many apricots does Allan have?

....

4. Brian can cycle five miles per hour. How far can Brian cycle in three hours?

....

5. Jennifer swims 11 laps every day. How many laps will Jennifer swim in seven days?

....

6. Ellen has 10 times more pears than Audrey. Audrey has four pears. How many pears does Ellen have?

....

7. Jackie's garden has eight rows of pumpkins. Each row has nine pumpkins. How many pumpkins does Jackie have in all?

....

1. If there are 10 avocados in each box and there are six boxes, how many avocados are there in total?

2. Jake has 10 times more peaches than Michele. Michele has two peaches. How many peaches does Jake have?

3. Brian can cycle eight miles per hour. How far can Brian cycle in seven hours?

4. Jackie's garden has 11 rows of pumpkins. Each row has five pumpkins. How many pumpkins does Jackie have in all?

5. Amy swims 11 laps every day. How many laps will Amy swim in three days?

6. If there are eight apples in each box and there are six boxes, how many apples are there in total?

7. Marcie's garden has eight rows of pumpkins. Each row has six pumpkins. How many pumpkins does Marcie have in all?

1. Brian swims eight laps every day. How many laps will Brian swim in two days?

   ....

2. Michele's garden has two rows of pumpkins. Each row has four pumpkins. How many pumpkins does Michele have in all?

   ....

3. Jennifer has 10 times more apricots than Amy. Amy has four apricots. How many apricots does Jennifer have?

   ....

4. If there are six plums in each box and there are seven boxes, how many plums are there in total?

   ....

5. Allan can cycle five miles per hour. How far can Allan cycle in seven hours?

   ....

6. Brian can cycle seven miles per hour. How far can Brian cycle in five hours?

   ....

7. If there are 12 bananas in each box and there are five boxes, how many bananas are there in total?

   ....

1. David can cycle 11 miles per hour. How far can David cycle in nine hours?

   ....

2. If there are eight apricots in each box and there are eight boxes, how many apricots are there in total?

   ....

3. Janet has 12 times more plums than Jackie. Jackie has four plums. How many plums does Janet have?

   ....

4. David swims four laps every day. How many laps will David swim in two days?

   ....

5. Ellen's garden has eight rows of pumpkins. Each row has seven pumpkins. How many pumpkins does Ellen have in all?

   ....

6. If there are 12 oranges in each box and there are five boxes, how many oranges are there in total?

   ....

7. Adam swims five laps every day. How many laps will Adam swim in three days?

   ....

1. Paul has two times more apricots than Steven. Steven has four apricots. How many apricots does Paul have?

    ....

2. Sharon's garden has nine rows of pumpkins. Each row has nine pumpkins. How many pumpkins does Sharon have in all?

    ....

3. Donald can cycle nine miles per hour. How far can Donald cycle in seven hours?

    ....

4. If there are six marbles in each box and there are eight boxes, how many marbles are there in total?

    ....

5. Sandra swims two laps every day. How many laps will Sandra swim in four days?

    ....

6. Jennifer's garden has 10 rows of pumpkins. Each row has seven pumpkins. How many pumpkins does Jennifer have in all?

    ....

7. If there are four peaches in each box and there are three boxes, how many peaches are there in total?

    ....

1. If there are 12 marbles in each box and there are four boxes, how many marbles are there in total?

2. Janet's garden has 10 rows of pumpkins. Each row has five pumpkins. How many pumpkins does Janet have in all?

3. Allan swims eight laps every day. How many laps will Allan swim in six days?

4. Donald has five times more oranges than David. David has six oranges. How many oranges does Donald have?

5. Allan can cycle 11 miles per hour. How far can Allan cycle in seven hours?

6. Ellen's garden has 10 rows of pumpkins. Each row has eight pumpkins. How many pumpkins does Ellen have in all?

7. Sandra has two times more avocados than Ellen. Ellen has eight avocados. How many avocados does Sandra have?

1. Amy has 11 times more peaches than Ellen. Ellen has two peaches. How many peaches does Amy have?

    ....

2. Jake can cycle 10 miles per hour. How far can Jake cycle in eight hours?

    ....

3. Donald swims five laps every day. How many laps will Donald swim in six days?

    ....

4. Sharon's garden has two rows of pumpkins. Each row has three pumpkins. How many pumpkins does Sharon have in all?

    ....

5. If there are five apples in each box and there are two boxes, how many apples are there in total?

    ....

6. Brian has seven times more balls than Steven. Steven has three balls. How many balls does Brian have?

    ....

7. Paul can cycle four miles per hour. How far can Paul cycle in four hours?

    ....

1. Amy has nine times more plums than Billy. Billy has eight plums. How many plums does Amy have?

....

2. Michele's garden has eight rows of pumpkins. Each row has two pumpkins. How many pumpkins does Michele have in all?

....

3. Brian can cycle seven miles per hour. How far can Brian cycle in seven hours?

....

4. Ellen swims seven laps every day. How many laps will Ellen swim in two days?

....

5. If there are six apples in each box and there are two boxes, how many apples are there in total?

....

6. If there are seven bananas in each box and there are six boxes, how many bananas are there in total?

....

7. Brian can cycle three miles per hour. How far can Brian cycle in six hours?

....

DIVISION

1. You have 30 bananas and want to share them equally with 6 people. How many bananas would each person get?

2. A box of marbles weighs 50 pounds. If one marbles weighs 10 pounds, how many marbles are there in the box?

3. Adam ordered 5 pizzas. The bill for the pizzas came to $20. What was the cost of each pizza?

4. How many 7 cm pieces of rope can you cut from a rope that is 28 cm long?

5. Audrey made 15 cookies for a bake sale. She put the cookies in bags, with 5 cookies in each bag. How many bags did she have for the bake sale?

6. Paul is reading a book with 25 pages. If Paul wants to read the same number of pages every day, how many pages would Paul have to read each day to finish in 5 days?

7. A box of apples weighs 3 pounds. If one apples weighs 1 pounds, how many apples are there in the box?

1. How many 5 cm pieces of rope can you cut from a rope that is 30 cm long?

   ....

2. A box of apples weighs 36 pounds. If one apples weighs 9 pounds, how many apples are there in the box?

   ....

3. Sandra made 35 cookies for a bake sale. She put the cookies in bags, with 5 cookies in each bag. How many bags did she have for the bake sale?

   ....

4. Janet ordered 5 pizzas. The bill for the pizzas came to $25. What was the cost of each pizza?

   ....

5. Brian is reading a book with 9 pages. If Brian wants to read the same number of pages every day, how many pages would Brian have to read each day to finish in 9 days?

   ....

6. You have 15 peaches and want to share them equally with 5 people. How many peaches would each person get?

   ....

7. Allan is reading a book with 16 pages. If Allan wants to read the same number of pages every day, how many pages would Allan have to read each day to finish in 4 days?

   ....

1. A box of plums weighs 90 pounds. If one plums weighs 10 pounds, how many plums are there in the box?

    ....

2. How many 6 cm pieces of rope can you cut from a rope that is 18 cm long?

    ....

3. Marcie made 8 cookies for a bake sale. She put the cookies in bags, with 4 cookies in each bag. How many bags did she have for the bake sale?

    ....

4. Donald is reading a book with 9 pages. If Donald wants to read the same number of pages every day, how many pages would Donald have to read each day to finish in 1 days?

    ....

5. You have 27 pears and want to share them equally with 3 people. How many pears would each person get?

    ....

6. Marcie ordered 8 pizzas. The bill for the pizzas came to $80. What was the cost of each pizza?

    ....

7. Michele made 16 cookies for a bake sale. She put the cookies in bags, with 2 cookies in each bag. How many bags did she have for the bake sale?

    ....

1. Ellen made 45 cookies for a bake sale. She put the cookies in bags, with 5 cookies in each bag. How many bags did she have for the bake sale?

   ....

2. Adam ordered 2 pizzas. The bill for the pizzas came to $10. What was the cost of each pizza?

   ....

3. A box of apples weighs 4 pounds. If one apples weighs 1 pounds, how many apples are there in the box?

   ....

4. You have 21 peaches and want to share them equally with 7 people. How many peaches would each person get?

   ....

5. How many 7 cm pieces of rope can you cut from a rope that is 49 cm long?

   ....

6. Steven is reading a book with 30 pages. If Steven wants to read the same number of pages every day, how many pages would Steven have to read each day to finish in 6 days?

   ....

7. A box of avocados weighs 63 pounds. If one avocados weighs 9 pounds, how many avocados are there in the box?

   ....

1. Marcie ordered 10 pizzas. The bill for the pizzas came to $50. What was the cost of each pizza?

   ....

2. Marcie made 4 cookies for a bake sale. She put the cookies in bags, with 1 cookies in each bag. How many bags did she have for the bake sale?

   ....

3. A box of apricots weighs 72 pounds. If one apricots weighs 8 pounds, how many apricots are there in the box?

   ....

4. You have 40 pears and want to share them equally with 8 people. How many pears would each person get?

   ....

5. How many 4 cm pieces of rope can you cut from a rope that is 40 cm long?

   ....

6. Brian is reading a book with 12 pages. If Brian wants to read the same number of pages every day, how many pages would Brian have to read each day to finish in 3 days?

   ....

7. How many 2 cm pieces of rope can you cut from a rope that is 8 cm long?

   ....

1. You have 48 balls and want to share them equally with 8 people. How many balls would each person get?

2. Jake is reading a book with 8 pages. If Jake wants to read the same number of pages every day, how many pages would Jake have to read each day to finish in 1 days?

3. How many 8 cm pieces of rope can you cut from a rope that is 72 cm long?

4. Jackie ordered 2 pizzas. The bill for the pizzas came to $10. What was the cost of each pizza?

5. A box of pears weighs 18 pounds. If one pears weighs 9 pounds, how many pears are there in the box?

6. Amy made 10 cookies for a bake sale. She put the cookies in bags, with 5 cookies in each bag. How many bags did she have for the bake sale?

7. How many 6 cm pieces of rope can you cut from a rope that is 42 cm long?

1. Janet made 4 cookies for a bake sale. She put the cookies in bags, with 1 cookies in each bag. How many bags did she have for the bake sale?

    ....

2. How many 5 cm pieces of rope can you cut from a rope that is 5 cm long?

    ....

3. Jake is reading a book with 4 pages. If Jake wants to read the same number of pages every day, how many pages would Jake have to read each day to finish in 4 days?

    ....

4. You have 40 balls and want to share them equally with 10 people. How many balls would each person get?

    ....

5. A box of apples weighs 36 pounds. If one apples weighs 6 pounds, how many apples are there in the box?

    ....

6. Ellen ordered 7 pizzas. The bill for the pizzas came to $7. What was the cost of each pizza?

    ....

7. How many 9 cm pieces of rope can you cut from a rope that is 72 cm long?

    ....

1. Ellen ordered 4 pizzas. The bill for the pizzas came to $36. What was the cost of each pizza?

2. Jackie made 3 cookies for a bake sale. She put the cookies in bags, with 1 cookies in each bag. How many bags did she have for the bake sale?

3. A box of apricots weighs 9 pounds. If one apricots weighs 9 pounds, how many apricots are there in the box?

4. Allan is reading a book with 18 pages. If Allan wants to read the same number of pages every day, how many pages would Allan have to read each day to finish in 6 days?

5. How many 7 cm pieces of rope can you cut from a rope that is 56 cm long?

6. You have 100 bananas and want to share them equally with 10 people. How many bananas would each person get?

7. Jackie made 36 cookies for a bake sale. She put the cookies in bags, with 6 cookies in each bag. How many bags did she have for the bake sale?

1. You have 35 oranges and want to share them equally with 5 people. How many oranges would each person get?

2. A box of avocados weighs 60 pounds. If one avocados weighs 10 pounds, how many avocados are there in the box?

3. Donald is reading a book with 28 pages. If Donald wants to read the same number of pages every day, how many pages would Donald have to read each day to finish in 7 days?

4. How many 6 cm pieces of rope can you cut from a rope that is 18 cm long?

5. Amy ordered 9 pizzas. The bill for the pizzas came to $18. What was the cost of each pizza?

6. Marcie made 12 cookies for a bake sale. She put the cookies in bags, with 4 cookies in each bag. How many bags did she have for the bake sale?

7. A box of marbles weighs 24 pounds. If one marbles weighs 3 pounds, how many marbles are there in the box?

1. Billy ordered 8 pizzas. The bill for the pizzas came to $56. What was the cost of each pizza?

2. You have 30 peaches and want to share them equally with 10 people. How many peaches would each person get?

3. Brian is reading a book with 18 pages. If Brian wants to read the same number of pages every day, how many pages would Brian have to read each day to finish in 3 days?

4. How many 4 cm pieces of rope can you cut from a rope that is 8 cm long?

5. Marcie made 72 cookies for a bake sale. She put the cookies in bags, with 8 cookies in each bag. How many bags did she have for the bake sale?

6. A box of apples weighs 40 pounds. If one apples weighs 8 pounds, how many apples are there in the box?

7. Marcie made 5 cookies for a bake sale. She put the cookies in bags, with 1 cookies in each bag. How many bags did she have for the bake sale?

SHOPPING PROBLEMS

hot dog = $1.50	cola = $1.00
order of French-fries = $1.00	ice cream cone = $1.00
hamburger = $2.50	milk shake = $2.00
deluxe cheeseburger = $4.00	taco = $2.50

1. .... What is the total cost of five ice cream cones, a milk shake, three tacos, an order of French-fries, and five deluxe cheeseburgers?

2. .... What is the total cost of four colas, four deluxe cheeseburgers, four orders of French-fries, two hamburgers, and an ice cream cone if there is a 20 percent discount?

3. .... Adam wants to buy four orders of French-fries, five milk shakes, a taco, and an ice cream cone. How much money will he need?

4. .... If Janet buys four tacos, five colas, three milk shakes, and five hamburgers, and if she had $40.00, how much money will she have left?

5. .... What is the total cost of five orders of French-fries, a taco, four hamburgers, and five milk shakes if there is a 20% discount?

6. .... Brian purchases a cola and three hamburgers. How much change will he get back from $20.00?

7. .... If Jackie buys a hot dog, two milk shakes, five colas, and an order of French-fries, and if she had $20.00, how much money will she have left?

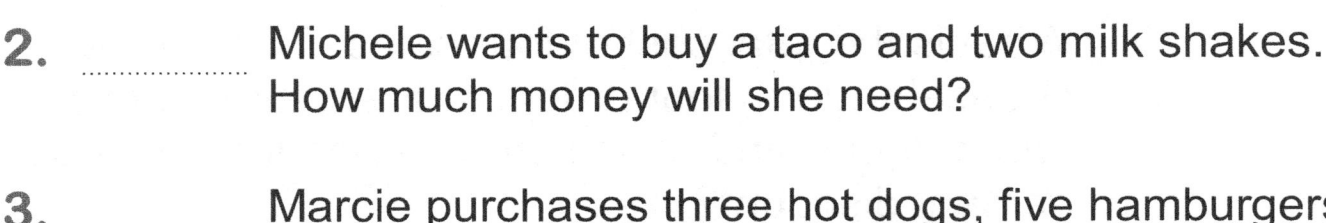

hot dog = $1.00   cola = $1.00
order of French-fries = $1.00   ice cream cone = $1.00
hamburger = $2.50   milk shake = $2.00
deluxe cheeseburger = $3.50   taco = $2.50

1. .... If David wanted to buy a hamburger, three colas, four tacos, and three ice cream cones, how much money would he need?

2. .... Michele wants to buy a taco and two milk shakes. How much money will she need?

3. .... Marcie purchases three hot dogs, five hamburgers, four colas, five orders of French-fries, and two milk shakes. How much money will she get back if she pays $40.00?

4. .... If Billy buys two hamburgers and three ice cream cones, how much change will he get back from $20.00?

5. .... Amy purchases three ice cream cones, five colas, and a milk shake. How much change will she get back from $15.00?

6. .... What is the total cost of two tacos, two milk shakes, and five orders of French-fries if there is a 20% discount?

7. .... What is the total cost of an ice cream cone, a hot dog, four hamburgers, a milk shake, and two deluxe cheeseburgers?

hot dog = $1.00	cola = $1.00
order of French-fries = $1.00	ice cream cone = $1.50
hamburger = $2.00	milk shake = $2.00
deluxe cheeseburger = $3.00	taco = $2.00

1. .... Billy wants to buy three colas, three milk shakes, and an ice cream cone. How much money will he need?

2. .... If Sandra buys an ice cream cone, five deluxe cheeseburgers, three colas, two orders of French-fries, and three hot dogs, what will her change be if she pays $30.00?

3. .... What is the total cost of two hamburgers and an order of French-fries?

4. .... Jackie purchases three colas, two hot dogs, and four deluxe cheeseburgers. If she had $30.00, how much money will she have left?

5. .... What is the total cost of an ice cream cone, five colas, four milk shakes, five hamburgers, and five tacos?

6. .... What is the total cost of five milk shakes?

7. .... What is the total cost of three hamburgers if there is a 20 percent discount?

hot dog = $1.00	cola = $1.00
order of French-fries = $1.50	ice cream cone = $1.50
hamburger = $2.00	milk shake = $3.00
deluxe cheeseburger = $3.00	taco = $2.00

1. _____ What is the total cost of five colas if there is a 20 percent discount?

2. _____ Marcie wants to buy three hot dogs, two tacos, and four hamburgers. How much will it cost her?

3. _____ Jackie purchases four hamburgers and two hot dogs. How much money will she get back if she pays $20.00?

4. _____ What is the total cost of three colas?

5. _____ If Brian buys five hamburgers, two milk shakes, two deluxe cheeseburgers, and five orders of French-fries, how much change will he get back from $40.00?

6. _____ What is the total cost of four tacos, a hamburger, three hot dogs, four deluxe cheeseburgers, and a cola if the items are on sale for 20 percent off the regular price?

7. _____ If Allan buys a hot dog, five orders of French-fries, and four ice cream cones, and if he had $20.00, how much money will he have left?

hot dog = $1.00	cola = $1.00
order of French-fries = $0.50	ice cream cone = $1.50
hamburger = $2.50	milk shake = $3.00
deluxe cheeseburger = $3.00	taco = $2.00

1. .... If Jennifer buys four tacos and three deluxe cheeseburgers, how much change will she get back from $20.00?

2. .... What is the total cost of five colas if there is a 20 percent discount?

3. .... What is the total cost of a cola, five hamburgers, three deluxe cheeseburgers, and five orders of French-fries if the items are on sale for 20 percent off the regular price?

4. .... What is the total cost of three colas, three milk shakes, two deluxe cheeseburgers, three tacos, and a hamburger?

5. .... Michele purchases three ice cream cones. How much change will she get back from $10.00?

6. .... Jake wants to buy two milk shakes, three hot dogs, and three orders of French-fries. How much will he have to pay?

7. .... What is the total cost of a cola and five tacos?

hot dog = $1.50	cola = $1.00
order of French-fries = $0.50	ice cream cone = $1.00
hamburger = $2.00	milk shake = $2.00
deluxe cheeseburger = $3.50	taco = $2.50

1. .... If Audrey buys three colas and four ice cream cones, what will her change be if she pays $20.00?

2. .... If Paul buys four orders of French-fries, five colas, and two milk shakes, what will his change be if he pays $15.00?

3. .... If Adam buys five tacos, five colas, four orders of French-fries, three milk shakes, and five hot dogs, how much change will he get back from $40.00?

4. .... Sandra wants to buy a hot dog and a cola. How much will she have to pay?

5. .... If Janet wanted to buy three milk shakes, four hamburgers, an order of French-fries, and two hot dogs, how much would it cost her?

6. .... What is the total cost of three deluxe cheeseburgers and five milk shakes?

7. .... What is the total cost of four orders of French-fries, four milk shakes, five colas, four tacos, and five ice cream cones?

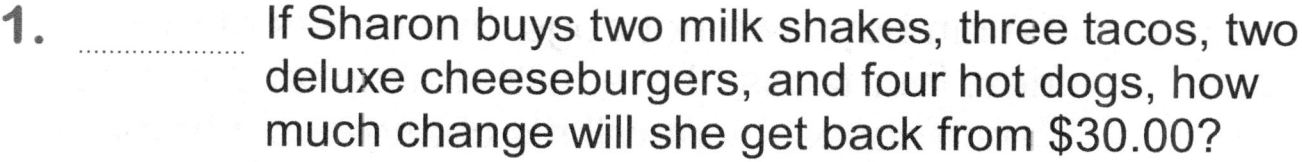

hot dog = $1.50   cola = $1.00
order of French-fries = $0.50   ice cream cone = $1.50
hamburger = $2.00   milk shake = $2.50
deluxe cheeseburger = $3.00   taco = $2.50

1. .... If Sharon buys two milk shakes, three tacos, two deluxe cheeseburgers, and four hot dogs, how much change will she get back from $30.00?

2. .... If Adam wanted to buy four orders of French-fries, how much would it cost him?

3. .... If Brian wanted to buy three colas, how much would it cost him?

4. .... What is the total cost of four hamburgers?

5. .... Ellen wants to buy an ice cream cone, a cola, three hot dogs, and a deluxe cheeseburger. How much will it cost her?

6. .... If Steven buys two orders of French-fries, three hamburgers, four deluxe cheeseburgers, and two colas, what will his change be if he pays $30.00?

7. .... If Janet wanted to buy four orders of French-fries and three milk shakes, how much would she have to pay?

hot dog = $1.50	cola = $1.00
order of French-fries = $0.50	ice cream cone = $1.50
hamburger = $2.00	milk shake = $2.50
deluxe cheeseburger = $4.00	taco = $2.00

1. .... If Allan buys two hot dogs, five orders of French-fries, four tacos, three milk shakes, and five ice cream cones, what will his change be if he pays $40.00?

2. .... Jackie purchases three hamburgers. If she had $20.00, how much money will she have left?

3. .... Paul purchases three ice cream cones, two deluxe cheeseburgers, and four hot dogs. How much money will he get back if he pays $30.00?

4. .... What is the total cost of three tacos and a cola?

5. .... Ellen wants to buy two orders of French-fries, a taco, a cola, and two ice cream cones. How much will she have to pay?

6. .... Michele wants to buy an order of French-fries, four hot dogs, four deluxe cheeseburgers, three colas, and an ice cream cone. How much will it cost her?

7. .... What is the total cost of five deluxe cheeseburgers?

hot dog = $1.00	cola = $1.00
order of French-fries = $0.50	ice cream cone = $1.50
hamburger = $2.50	milk shake = $2.00
deluxe cheeseburger = $3.50	taco = $2.00

1. .... If Steven buys three hot dogs, and if he had $10.00, how much money will he have left?

2. .... Allan wants to buy an order of French-fries, two colas, four hamburgers, and three tacos. How much will it cost him?

3. .... Marcie purchases five tacos and a cola. If she had $15.00, how much money will she have left?

4. .... Billy purchases two hamburgers, a deluxe cheeseburger, and four ice cream cones. How much change will he get back from $20.00?

5. .... What is the total cost of a hot dog, five milk shakes, two colas, four tacos, and a hamburger?

6. .... If Audrey wanted to buy five orders of French-fries and four tacos, how much money would she need?

7. .... If Michele wanted to buy five hamburgers and five tacos, how much money would she need?

hot dog = $1.50	cola = $1.00
order of French-fries = $1.00	ice cream cone = $1.00
hamburger = $2.00	milk shake = $2.00
deluxe cheeseburger = $3.50	taco = $2.00

1. .... Adam wants to buy two ice cream cones, two hamburgers, and five tacos. How much will it cost him?

2. .... Allan purchases three orders of French-fries. How much money will he get back if he pays $10.00?

3. .... What is the total cost of five deluxe cheeseburgers?

4. .... What is the total cost of three deluxe cheeseburgers and two orders of French-fries?

5. .... Paul wants to buy three hamburgers, a taco, five ice cream cones, and four deluxe cheeseburgers. How much will he have to pay?

6. .... What is the total cost of two milk shakes?

7. .... If Billy wanted to buy an ice cream cone, four milk shakes, three hamburgers, and five colas, how much would it cost him?

ANSWERS

1. 3 red peaches and 19 green peaches are in the basket. How many peaches are in the basket?

   22

2. 18 oranges were in the basket. More oranges were added to the basket. Now there are 19 oranges. How many oranges were added to the basket?

   1

3. Marcie has 12 more bananas than Audrey. Audrey has 6 bananas. How many bananas does Marcie have?

   18

4. Paul has 7 plums and Jake has 2 plums. How many plums do Paul and Jake have together?

   9

5. 24 pears were in the basket. 4 are red and the rest are green. How many pears are green?

   20

6. 4 marbles were in the basket. More marbles were added to the basket. Now there are 21 marbles. How many marbles were added to the basket?

   17

7. Marcie has 12 more avocados than Jackie. Jackie has 8 avocados. How many avocados does Marcie have?

   20

1. 2 red marbles and 5 green marbles are in the basket. How many marbles are in the basket?

   7

2. 28 apples were in the basket. 17 are red and the rest are green. How many apples are green?

   11

3. 16 apricots were in the basket. More apricots were added to the basket. Now there are 32 apricots. How many apricots were added to the basket?

   16

4. Marcie has 8 more peaches than Jackie. Jackie has 16 peaches. How many peaches does Marcie have?

   24

5. Adam has 6 avocados and Brian has 20 avocados. How many avocados do Adam and Brian have together?

   26

6. 6 pears were in the basket. More pears were added to the basket. Now there are 26 pears. How many pears were added to the basket?

   20

7. Jake has 7 balls and Billy has 5 balls. How many balls do Jake and Billy have together?

   12

1. 4 red apricots and 14 green apricots are in the basket. How many apricots are in the basket?

   18

2. 17 apples were in the basket. 7 are red and the rest are green. How many apples are green?

   10

3. Jennifer has 13 more avocados than Marin. Marin has 12 avocados. How many avocados does Jennifer have?

   25

4. 6 peaches were in the basket. More peaches were added to the basket. Now there are 11 peaches. How many peaches were added to the basket?

   5

5. Allan has 5 oranges and Billy has 11 oranges. How many oranges do Allan and Billy have together?

   16

6. Audrey has 20 more marbles than Marcie. Marcie has 5 marbles. How many marbles does Audrey have?

   25

7. 8 red plums and 15 green plums are in the basket. How many plums are in the basket?

   23

1. 17 red pears and 15 green pears are in the basket. How many pears are in the basket?

   32

2. Steven has 2 avocados and Jake has 10 avocados. How many avocados do Steven and Jake have together?

   12

3. Ellen has 17 more marbles than Jackie. Jackie has 15 marbles. How many marbles does Ellen have?

   32

4. 10 plums were in the basket. More plums were added to the basket. Now there are 16 plums. How many plums were added to the basket?

   6

5. 10 bananas were in the basket. 9 are red and the rest are green. How many bananas are green?

   1

6. 5 balls were in the basket. More balls were added to the basket. Now there are 7 balls. How many balls were added to the basket?

   2

7. 19 red oranges and 20 green oranges are in the basket. How many oranges are in the basket?

   39

1. 12 peaches were in the basket. 1 is red and the rest are green. How many peaches are green?

11

2. 12 red avocados and 3 green avocados are in the basket. How many avocados are in the basket?

15

3. Marin has 15 more marbles than Jackie. Jackie has 10 marbles. How many marbles does Marin have?

25

4. 12 apricots were in the basket. More apricots were added to the basket. Now there are 31 apricots. How many apricots were added to the basket?

19

5. Donald has 15 oranges and Billy has 16 oranges. How many oranges do Donald and Billy have together?

31

6. Brian has 18 pears and Donald has 2 pears. How many pears do Brian and Donald have together?

20

7. 2 bananas were in the basket. More bananas were added to the basket. Now there are 18 bananas. How many bananas were added to the basket?

16

1. 8 red bananas and 9 green bananas are in the basket. How many bananas are in the basket?

17

2. 21 balls were in the basket. 17 are red and the rest are green. How many balls are green?

4

3. Brian has 4 apples and Paul has 16 apples. How many apples do Brian and Paul have together?

20

4. 6 pears were in the basket. More pears were added to the basket. Now there are 8 pears. How many pears were added to the basket?

2

5. Jennifer has 2 more avocados than Janet. Janet has 1 avocado. How many avocados does Jennifer have?

3

6. 7 red plums and 16 green plums are in the basket. How many plums are in the basket?

23

7. 19 oranges were in the basket. 11 are red and the rest are green. How many oranges are green?

8

1. 17 pears were in the basket. More pears were added to the basket. Now there are 23 pears. How many pears were added to the basket?

6

2. Paul has 1 ball and Allan has 20 balls. How many balls do Paul and Allan have together?

21

3. Jackie has 18 more bananas than Jennifer. Jennifer has 10 bananas. How many bananas does Jackie have?

28

4. 10 avocados were in the basket. 7 are red and the rest are green. How many avocados are green?

3

5. 5 red peaches and 5 green peaches are in the basket. How many peaches are in the basket?

10

6. 15 red apricots and 11 green apricots are in the basket. How many apricots are in the basket?

26

7. 18 oranges were in the basket. 6 are red and the rest are green. How many oranges are green?

12

1. 16 pears were in the basket. More pears were added to the basket. Now there are 24 pears. How many pears were added to the basket?

8

2. 6 red plums and 12 green plums are in the basket. How many plums are in the basket?

18

3. Allan has 5 peaches and Donald has 7 peaches. How many peaches do Allan and Donald have together?

12

4. Amy has 2 more apples than Audrey. Audrey has 8 apples. How many apples does Amy have?

10

5. 23 avocados were in the basket. 8 are red and the rest are green. How many avocados are green?

15

6. Billy has 11 marbles and David has 1 marble. How many marbles do Billy and David have together?

12

7. 7 balls were in the basket. 5 are red and the rest are green. How many balls are green?

2

1. Donald has 4 bananas and Steven has 13 bananas. How many bananas do Donald and Steven have together?

   17

2. 3 red peaches and 3 green peaches are in the basket. How many peaches are in the basket?

   6

3. 8 oranges were in the basket. More oranges were added to the basket. Now there are 20 oranges. How many oranges were added to the basket?

   12

4. Jennifer has 10 more avocados than Marin. Marin has 14 avocados. How many avocados does Jennifer have?

   24

5. 17 plums were in the basket. 11 are red and the rest are green. How many plums are green?

   6

6. 17 red balls and 12 green balls are in the basket. How many balls are in the basket?

   29

7. Marcie has 15 more pears than Jackie. Jackie has 4 pears. How many pears does Marcie have?

   19

1. 38 peaches were in the basket. 19 are red and the rest are green. How many peaches are green?

   19

2. 19 red plums and 18 green plums are in the basket. How many plums are in the basket?

   37

3. Jackie has 14 more apples than Janet. Janet has 16 apples. How many apples does Jackie have?

   30

4. Donald has 10 bananas and Allan has 8 bananas. How many bananas do Donald and Allan have together?

   18

5. 1 ball was in the basket. More balls were added to the basket. Now there are 10 balls. How many balls were added to the basket?

   9

6. Steven has 19 pears and Allan has 6 pears. How many pears do Steven and Allan have together?

   25

7. 10 red apricots and 4 green apricots are in the basket. How many apricots are in the basket?

   14

1. 8 apricots are in the basket. 4 apricots are taken out of the basket. How many apricots are in the basket now?

   4

2. Some oranges were in the basket. 14 oranges were taken from the basket. Now there are 0 oranges. How many oranges were in the basket before some of the oranges were taken?

   14

3. Donald has 5 peaches. David has 6 peaches. How many more peaches does David have than Donald?

   1

4. 19 bananas were in the basket. Some of the bananas were removed from the basket. Now there are 7 bananas. How many bananas were removed from the basket?

   12

5. Audrey has 7 fewer balls than Michele. Michele has 18 balls. How many balls does Audrey have?

   11

6. 10 marbles are in the basket. 9 are red and the rest are green. How many marbles are green?

   1

7. David has 2 pears. Allan has 19 pears. How many more pears does Allan have than David?

   17

1. Donald has 13 bananas. Billy has 19 bananas. How many more bananas does Billy have than Donald?

   6

2. Some oranges were in the basket. 2 oranges were taken from the basket. Now there are 5 oranges. How many oranges were in the basket before some of the oranges were taken?

   7

3. 14 pears are in the basket. 4 pears are taken out of the basket. How many pears are in the basket now?

   10

4. 6 peaches are in the basket. 3 are red and the rest are green. How many peaches are green?

   3

5. Marcie has 2 fewer marbles than Jennifer. Jennifer has 6 marbles. How many marbles does Marcie have?

   4

6. 7 apples were in the basket. Some of the apples were removed from the basket. Now there are 6 apples. How many apples were removed from the basket?

   1

7. Some plums were in the basket. 6 plums were taken from the basket. Now there are 0 plums. How many plums were in the basket before some of the plums were taken?

   6

1. 13 apricots were in the basket. Some of the apricots were removed from the basket. Now there are 8 apricots. How many apricots were removed from the basket?

    5

2. Amy has 2 fewer oranges than Michele. Michele has 10 oranges. How many oranges does Amy have?

    8

3. Steven has 3 balls. Adam has 12 balls. How many more balls does Adam have than Steven?

    9

4. Some marbles were in the basket. 8 marbles were taken from the basket. Now there are 7 marbles. How many marbles were in the basket before some of the marbles were taken?

    15

5. 7 avocados are in the basket. 5 avocados are taken out of the basket. How many avocados are in the basket now?

    2

6. 15 pears are in the basket. 8 are red and the rest are green. How many pears are green?

    7

7. Michele has 0 fewer apples than Marin. Marin has 2 apples. How many apples does Michele have?

    2

1. David has 12 apples. Donald has 13 apples. How many more apples does Donald have than David?

    1

2. 6 marbles are in the basket. 1 is red and the rest are green. How many marbles are green?

    5

3. 4 balls are in the basket. 1 ball is taken out of the basket. How many balls are in the basket now?

    3

4. Some avocados were in the basket. 2 avocados were taken from the basket. Now there are 4 avocados. How many avocados were in the basket before some of the avocados were taken?

    6

5. 1 banana was in the basket. Some of the bananas were removed from the basket. Now there are 0 bananas. How many bananas were removed from the basket?

    1

6. Michele has 3 fewer peaches than Marcie. Marcie has 18 peaches. How many peaches does Michele have?

    15

7. 20 plums were in the basket. Some of the plums were removed from the basket. Now there are 19 plums. How many plums were removed from the basket?

    1

1. 9 oranges were in the basket. Some of the oranges were removed from the basket. Now there is 1 orange. How many oranges were removed from the basket?

    8

2. Some bananas were in the basket. 3 bananas were taken from the basket. Now there are 3 bananas. How many bananas were in the basket before some of the bananas were taken?

    6

3. Audrey has 4 fewer avocados than Sandra. Sandra has 16 avocados. How many avocados does Audrey have?

    12

4. 18 plums are in the basket. 15 plums are taken out of the basket. How many plums are in the basket now?

    3

5. Brian has 4 marbles. Allan has 20 marbles. How many more marbles does Allan have than Brian?

    16

6. 14 apples are in the basket. 14 are red and the rest are green. How many apples are green?

    0

7. 17 apricots are in the basket. 2 are red and the rest are green. How many apricots are green?

    15

1. Some pears were in the basket. 5 pears were taken from the basket. Now there is 1 pear. How many pears were in the basket before some of the pears were taken?

    6

2. 8 apricots were in the basket. Some of the apricots were removed from the basket. Now there are 3 apricots. How many apricots were removed from the basket?

    5

3. 20 bananas are in the basket. 19 are red and the rest are green. How many bananas are green?

    1

4. Marcie has 0 fewer marbles than Marin. Marin has 13 marbles. How many marbles does Marcie have?

    13

5. Billy has 5 avocados. Steven has 15 avocados. How many more avocados does Steven have than Billy?

    10

6. 20 balls are in the basket. 12 balls are taken out of the basket. How many balls are in the basket now?

    8

7. Paul has 8 peaches. Donald has 8 peaches. How many more peaches does Donald have than Paul?

    0

1. 2 pears were in the basket. Some of the pears were removed from the basket. Now there is 1 pear. How many pears were removed from the basket?

    1

2. Amy has 11 fewer marbles than Sharon. Sharon has 12 marbles. How many marbles does Amy have?

    1

3. Some peaches were in the basket. 6 peaches were taken from the basket. Now there are 5 peaches. How many peaches were in the basket before some of the peaches were taken?

    11

4. 15 oranges are in the basket. 12 oranges are taken out of the basket. How many oranges are in the basket now?

    3

5. 19 balls are in the basket. 11 are red and the rest are green. How many balls are green?

    8

6. Steven has 2 avocados. Allan has 5 avocados. How many more avocados does Allan have than Steven?

    3

7. 9 bananas are in the basket. 6 bananas are taken out of the basket. How many bananas are in the basket now?

    3

1. 17 apples were in the basket. Some of the apples were removed from the basket. Now there are 2 apples. How many apples were removed from the basket?

    15

2. 14 bananas are in the basket. 13 are red and the rest are green. How many bananas are green?

    1

3. Some apricots were in the basket. 10 apricots were taken from the basket. Now there are 4 apricots. How many apricots were in the basket before some of the apricots were taken?

    14

4. 4 peaches are in the basket. 4 peaches are taken out of the basket. How many peaches are in the basket now?

    0

5. Billy has 18 plums. David has 20 plums. How many more plums does David have than Billy?

    2

6. Audrey has 1 fewer marble than Jennifer. Jennifer has 12 marbles. How many marbles does Audrey have?

    11

7. 17 balls are in the basket. 16 are red and the rest are green. How many balls are green?

    1

1. Some apples were in the basket. 3 apples were taken from the basket. Now there is 1 apple. How many apples were in the basket before some of the apples were taken?

    4

2. 4 oranges were in the basket. Some of the oranges were removed from the basket. Now there are 3 oranges. How many oranges were removed from the basket?

    1

3. 19 pears are in the basket. 14 are red and the rest are green. How many pears are green?

    5

4. 4 marbles are in the basket. 4 marbles are taken out of the basket. How many marbles are in the basket now?

    0

5. Paul has 1 plum. Steven has 18 plums. How many more plums does Steven have than Paul?

    17

6. Audrey has 2 fewer avocados than Sharon. Sharon has 11 avocados. How many avocados does Audrey have?

    9

7. Billy has 8 bananas. David has 10 bananas. How many more bananas does David have than Billy?

    2

1. 14 plums are in the basket. 1 is red and the rest are green. How many plums are green?

    13

2. 11 pears were in the basket. Some of the pears were removed from the basket. Now there are 0 pears. How many pears were removed from the basket?

    11

3. 20 balls are in the basket. 2 balls are taken out of the basket. How many balls are in the basket now?

    18

4. Jackie has 6 fewer apples than Jennifer. Jennifer has 9 apples. How many apples does Jackie have?

    3

5. Brian has 9 oranges. Jake has 17 oranges. How many more oranges does Jake have than Brian?

    8

6. Some peaches were in the basket. 3 peaches were taken from the basket. Now there are 6 peaches. How many peaches were in the basket before some of the peaches were taken?

    9

7. 2 bananas were in the basket. Some of the bananas were removed from the basket. Now there are 0 bananas. How many bananas were removed from the basket?

    2

1. If there are two oranges in each box and there are four boxes, how many oranges are there in total?

   8

2. Sharon's garden has six rows of pumpkins. Each row has eight pumpkins. How many pumpkins does Sharon have in all?

   48

3. Steven swims three laps every day. How many laps will Steven swim in three days?

   9

4. Allan can cycle eight miles per hour. How far can Allan cycle in five hours?

   40

5. Janet has 10 times more plums than Amy. Amy has seven plums. How many plums does Janet have?

   70

6. Jennifer's garden has nine rows of pumpkins. Each row has three pumpkins. How many pumpkins does Jennifer have in all?

   27

7. Steven can cycle 11 miles per hour. How far can Steven cycle in seven hours?

   77

1. Paul swims 12 laps every day. How many laps will Paul swim in two days?

   24

2. Jennifer's garden has nine rows of pumpkins. Each row has six pumpkins. How many pumpkins does Jennifer have in all?

   54

3. Marcie has 10 times more peaches than Ellen. Ellen has three peaches. How many peaches does Marcie have?

   30

4. If there are six balls in each box and there are three boxes, how many balls are there in total?

   18

5. Jake can cycle eight miles per hour. How far can Jake cycle in eight hours?

   64

6. If there are three oranges in each box and there are six boxes, how many oranges are there in total?

   18

7. Michele swims eight laps every day. How many laps will Michele swim in nine days?

   72

1. If there are seven oranges in each box and there are three boxes, how many oranges are there in total?

   21

2. Amy's garden has 11 rows of pumpkins. Each row has four pumpkins. How many pumpkins does Amy have in all?

   44

3. Allan has two times more apricots than Janet. Janet has five apricots. How many apricots does Allan have?

   10

4. Brian can cycle five miles per hour. How far can Brian cycle in three hours?

   15

5. Jennifer swims 11 laps every day. How many laps will Jennifer swim in seven days?

   77

6. Ellen has 10 times more pears than Audrey. Audrey has four pears. How many pears does Ellen have?

   40

7. Jackie's garden has eight rows of pumpkins. Each row has nine pumpkins. How many pumpkins does Jackie have in all?

   72

1. If there are 10 avocados in each box and there are six boxes, how many avocados are there in total?

   60

2. Jake has 10 times more peaches than Michele. Michele has two peaches. How many peaches does Jake have?

   20

3. Brian can cycle eight miles per hour. How far can Brian cycle in seven hours?

   56

4. Jackie's garden has 11 rows of pumpkins. Each row has five pumpkins. How many pumpkins does Jackie have in all?

   55

5. Amy swims 11 laps every day. How many laps will Amy swim in three days?

   33

6. If there are eight apples in each box and there are six boxes, how many apples are there in total?

   48

7. Marcie's garden has eight rows of pumpkins. Each row has six pumpkins. How many pumpkins does Marcie have in all?

   48

1. Brian swims eight laps every day. How many laps will Brian swim in two days?

   16

2. Michele's garden has two rows of pumpkins. Each row has four pumpkins. How many pumpkins does Michele have in all?

   8

3. Jennifer has 10 times more apricots than Amy. Amy has four apricots. How many apricots does Jennifer have?

   40

4. If there are six plums in each box and there are seven boxes, how many plums are there in total?

   42

5. Allan can cycle five miles per hour. How far can Allan cycle in seven hours?

   35

6. Brian can cycle seven miles per hour. How far can Brian cycle in five hours?

   35

7. If there are 12 bananas in each box and there are five boxes, how many bananas are there in total?

   60

1. David can cycle 11 miles per hour. How far can David cycle in nine hours?

   99

2. If there are eight apricots in each box and there are eight boxes, how many apricots are there in total?

   64

3. Janet has 12 times more plums than Jackie. Jackie has four plums. How many plums does Janet have?

   48

4. David swims four laps every day. How many laps will David swim in two days?

   8

5. Ellen's garden has eight rows of pumpkins. Each row has seven pumpkins. How many pumpkins does Ellen have in all?

   56

6. If there are 12 oranges in each box and there are five boxes, how many oranges are there in total?

   60

7. Adam swims five laps every day. How many laps will Adam swim in three days?

   15

1. Paul has two times more apricots than Steven. Steven has four apricots. How many apricots does Paul have?

   8

2. Sharon's garden has nine rows of pumpkins. Each row has nine pumpkins. How many pumpkins does Sharon have in all?

   81

3. Donald can cycle nine miles per hour. How far can Donald cycle in seven hours?

   63

4. If there are six marbles in each box and there are eight boxes, how many marbles are there in total?

   48

5. Sandra swims two laps every day. How many laps will Sandra swim in four days?

   8

6. Jennifer's garden has 10 rows of pumpkins. Each row has seven pumpkins. How many pumpkins does Jennifer have in all?

   70

7. If there are four peaches in each box and there are three boxes, how many peaches are there in total?

   12

1. If there are 12 marbles in each box and there are four boxes, how many marbles are there in total?

   48

2. Janet's garden has 10 rows of pumpkins. Each row has five pumpkins. How many pumpkins does Janet have in all?

   50

3. Allan swims eight laps every day. How many laps will Allan swim in six days?

   48

4. Donald has five times more oranges than David. David has six oranges. How many oranges does Donald have?

   30

5. Allan can cycle 11 miles per hour. How far can Allan cycle in seven hours?

   77

6. Ellen's garden has 10 rows of pumpkins. Each row has eight pumpkins. How many pumpkins does Ellen have in all?

   80

7. Sandra has two times more avocados than Ellen. Ellen has eight avocados. How many avocados does Sandra have?

   16

1. Amy has 11 times more peaches than Ellen. Ellen has two peaches. How many peaches does Amy have?

   22

2. Jake can cycle 10 miles per hour. How far can Jake cycle in eight hours?

   80

3. Donald swims five laps every day. How many laps will Donald swim in six days?

   30

4. Sharon's garden has two rows of pumpkins. Each row has three pumpkins. How many pumpkins does Sharon have in all?

   6

5. If there are five apples in each box and there are two boxes, how many apples are there in total?

   10

6. Brian has seven times more balls than Steven. Steven has three balls. How many balls does Brian have?

   21

7. Paul can cycle four miles per hour. How far can Paul cycle in four hours?

   16

1. Amy has nine times more plums than Billy. Billy has eight plums. How many plums does Amy have?

   72

2. Michele's garden has eight rows of pumpkins. Each row has two pumpkins. How many pumpkins does Michele have in all?

   16

3. Brian can cycle seven miles per hour. How far can Brian cycle in seven hours?

   49

4. Ellen swims seven laps every day. How many laps will Ellen swim in two days?

   14

5. If there are six apples in each box and there are two boxes, how many apples are there in total?

   12

6. If there are seven bananas in each box and there are six boxes, how many bananas are there in total?

   42

7. Brian can cycle three miles per hour. How far can Brian cycle in six hours?

   18

1. You have 30 bananas and want to share them equally with 6 people. How many bananas would each person get?

   5

2. A box of marbles weighs 50 pounds. If one marbles weighs 10 pounds, how many marbles are there in the box?

   5

3. Adam ordered 5 pizzas. The bill for the pizzas came to $20. What was the cost of each pizza?

   4

4. How many 7 cm pieces of rope can you cut from a rope that is 28 cm long?

   4

5. Audrey made 15 cookies for a bake sale. She put the cookies in bags, with 5 cookies in each bag. How many bags did she have for the bake sale?

   3

6. Paul is reading a book with 25 pages. If Paul wants to read the same number of pages every day, how many pages would Paul have to read each day to finish in 5 days?

   5

7. A box of apples weighs 3 pounds. If one apples weighs 1 pounds, how many apples are there in the box?

   3

1. How many 5 cm pieces of rope can you cut from a rope that is 30 cm long?

   6

2. A box of apples weighs 36 pounds. If one apples weighs 9 pounds, how many apples are there in the box?

   4

3. Sandra made 35 cookies for a bake sale. She put the cookies in bags, with 5 cookies in each bag. How many bags did she have for the bake sale?

   7

4. Janet ordered 5 pizzas. The bill for the pizzas came to $25. What was the cost of each pizza?

   5

5. Brian is reading a book with 9 pages. If Brian wants to read the same number of pages every day, how many pages would Brian have to read each day to finish in 9 days?

   1

6. You have 15 peaches and want to share them equally with 5 people. How many peaches would each person get?

   3

7. Allan is reading a book with 16 pages. If Allan wants to read the same number of pages every day, how many pages would Allan have to read each day to finish in 4 days?

   4

1. A box of plums weighs 90 pounds. If one plums weighs 10 pounds, how many plums are there in the box?

   9

2. How many 6 cm pieces of rope can you cut from a rope that is 18 cm long?

   3

3. Marcie made 8 cookies for a bake sale. She put the cookies in bags, with 4 cookies in each bag. How many bags did she have for the bake sale?

   2

4. Donald is reading a book with 9 pages. If Donald wants to read the same number of pages every day, how many pages would Donald have to read each day to finish in 1 days?

   9

5. You have 27 pears and want to share them equally with 3 people. How many pears would each person get?

   9

6. Marcie ordered 8 pizzas. The bill for the pizzas came to $80. What was the cost of each pizza?

   10

7. Michele made 16 cookies for a bake sale. She put the cookies in bags, with 2 cookies in each bag. How many bags did she have for the bake sale?

   8

1. Ellen made 45 cookies for a bake sale. She put the cookies in bags, with 5 cookies in each bag. How many bags did she have for the bake sale?

   9

2. Adam ordered 2 pizzas. The bill for the pizzas came to $10. What was the cost of each pizza?

   5

3. A box of apples weighs 4 pounds. If one apples weighs 1 pounds, how many apples are there in the box?

   4

4. You have 21 peaches and want to share them equally with 7 people. How many peaches would each person get?

   3

5. How many 7 cm pieces of rope can you cut from a rope that is 49 cm long?

   7

6. Steven is reading a book with 30 pages. If Steven wants to read the same number of pages every day, how many pages would Steven have to read each day to finish in 6 days?

   5

7. A box of avocados weighs 63 pounds. If one avocados weighs 9 pounds, how many avocados are there in the box?

   7

1. Marcie ordered 10 pizzas. The bill for the pizzas came to $50. What was the cost of each pizza?

   5

2. Marcie made 4 cookies for a bake sale. She put the cookies in bags, with 1 cookies in each bag. How many bags did she have for the bake sale?

   4

3. A box of apricots weighs 72 pounds. If one apricots weighs 8 pounds, how many apricots are there in the box?

   9

4. You have 40 pears and want to share them equally with 8 people. How many pears would each person get?

   5

5. How many 4 cm pieces of rope can you cut from a rope that is 40 cm long?

   10

6. Brian is reading a book with 12 pages. If Brian wants to read the same number of pages every day, how many pages would Brian have to read each day to finish in 3 days?

   4

7. How many 2 cm pieces of rope can you cut from a rope that is 8 cm long?

   4

1. You have 48 balls and want to share them equally with 8 people. How many balls would each person get?

   6

2. Jake is reading a book with 8 pages. If Jake wants to read the same number of pages every day, how many pages would Jake have to read each day to finish in 1 days?

   8

3. How many 8 cm pieces of rope can you cut from a rope that is 72 cm long?

   9

4. Jackie ordered 2 pizzas. The bill for the pizzas came to $10. What was the cost of each pizza?

   5

5. A box of pears weighs 18 pounds. If one pears weighs 9 pounds, how many pears are there in the box?

   2

6. Amy made 10 cookies for a bake sale. She put the cookies in bags, with 5 cookies in each bag. How many bags did she have for the bake sale?

   2

7. How many 6 cm pieces of rope can you cut from a rope that is 42 cm long?

   7

1. Janet made 4 cookies for a bake sale. She put the cookies in bags, with 1 cookies in each bag. How many bags did she have for the bake sale?

   4

2. How many 5 cm pieces of rope can you cut from a rope that is 5 cm long?

   1

3. Jake is reading a book with 4 pages. If Jake wants to read the same number of pages every day, how many pages would Jake have to read each day to finish in 4 days?

   1

4. You have 40 balls and want to share them equally with 10 people. How many balls would each person get?

   4

5. A box of apples weighs 36 pounds. If one apples weighs 6 pounds, how many apples are there in the box?

   6

6. Ellen ordered 7 pizzas. The bill for the pizzas came to $7. What was the cost of each pizza?

   1

7. How many 9 cm pieces of rope can you cut from a rope that is 72 cm long?

   8

1. Ellen ordered 4 pizzas. The bill for the pizzas came to $36. What was the cost of each pizza?

   9

2. Jackie made 3 cookies for a bake sale. She put the cookies in bags, with 1 cookies in each bag. How many bags did she have for the bake sale?

   3

3. A box of apricots weighs 9 pounds. If one apricots weighs 9 pounds, how many apricots are there in the box?

   1

4. Allan is reading a book with 18 pages. If Allan wants to read the same number of pages every day, how many pages would Allan have to read each day to finish in 6 days?

   3

5. How many 7 cm pieces of rope can you cut from a rope that is 56 cm long?

   8

6. You have 100 bananas and want to share them equally with 10 people. How many bananas would each person get?

   10

7. Jackie made 36 cookies for a bake sale. She put the cookies in bags, with 6 cookies in each bag. How many bags did she have for the bake sale?

   6

1. You have 35 oranges and want to share them equally with 5 people. How many oranges would each person get?

   7

2. A box of avocados weighs 60 pounds. If one avocados weighs 10 pounds, how many avocados are there in the box?

   6

3. Donald is reading a book with 28 pages. If Donald wants to read the same number of pages every day, how many pages would Donald have to read each day to finish in 7 days?

   4

4. How many 6 cm pieces of rope can you cut from a rope that is 18 cm long?

   3

5. Amy ordered 9 pizzas. The bill for the pizzas came to $18. What was the cost of each pizza?

   2

6. Marcie made 12 cookies for a bake sale. She put the cookies in bags, with 4 cookies in each bag. How many bags did she have for the bake sale?

   3

7. A box of marbles weighs 24 pounds. If one marbles weighs 3 pounds, how many marbles are there in the box?

   8

1. Billy ordered 8 pizzas. The bill for the pizzas came to $56. What was the cost of each pizza?

   7

2. You have 30 peaches and want to share them equally with 10 people. How many peaches would each person get?

   3

3. Brian is reading a book with 18 pages. If Brian wants to read the same number of pages every day, how many pages would Brian have to read each day to finish in 3 days?

   6

4. How many 4 cm pieces of rope can you cut from a rope that is 8 cm long?

   2

5. Marcie made 72 cookies for a bake sale. She put the cookies in bags, with 8 cookies in each bag. How many bags did she have for the bake sale?

   9

6. A box of apples weighs 40 pounds. If one apples weighs 8 pounds, how many apples are there in the box?

   5

7. Marcie made 5 cookies for a bake sale. She put the cookies in bags, with 1 cookies in each bag. How many bags did she have for the bake sale?

   5

Page 41

hot dog = $1.50	cola = $1.00
order of French-fries = $1.00	ice cream cone = $1.00
hamburger = $2.50	milk shake = $2.00
deluxe cheeseburger = $4.00	taco = $2.50

1. $35.50 — What is the total cost of five ice cream cones, a milk shake, three tacos, an order of French-fries, and five deluxe cheeseburgers?

2. $24.00 — What is the total cost of four colas, four deluxe cheeseburgers, four orders of French-fries, two hamburgers, and an ice cream cone if there is a 20 percent discount?

3. $17.50 — Adam wants to buy four orders of French-fries, five milk shakes, a taco, and an ice cream cone. How much money will he need?

4. $6.50 — If Janet buys four tacos, five colas, three milk shakes, and five hamburgers, and if she had $40.00, how much money will she have left?

5. $22.00 — What is the total cost of five orders of French-fries, a taco, four hamburgers, and five milk shakes if there is a 20% discount?

6. $11.50 — Brian purchases a cola and three hamburgers. How much change will he get back from $20.00?

7. $8.50 — If Jackie buys a hot dog, two milk shakes, five colas, and an order of French-fries, and if she had $20.00, how much money will she have left?

Page 42

hot dog = $1.00	cola = $1.00
order of French-fries = $1.00	ice cream cone = $1.00
hamburger = $2.50	milk shake = $2.00
deluxe cheeseburger = $3.50	taco = $2.50

1. $18.50 — If David wanted to buy a hamburger, three colas, four tacos, and three ice cream cones, how much money would he need?

2. $6.50 — Michele wants to buy a taco and two milk shakes. How much money will she need?

3. $11.50 — Marcie purchases three hot dogs, five hamburgers, four colas, five orders of French-fries, and two milk shakes. How much money will she get back if she pays $40.00?

4. $12.00 — If Billy buys two hamburgers and three ice cream cones, how much change will he get back from $20.00?

5. $5.00 — Amy purchases three ice cream cones, five colas, and a milk shake. How much change will she get back from $15.00?

6. $11.20 — What is the total cost of two tacos, two milk shakes, and five orders of French-fries if there is a 20% discount?

7. $21.00 — What is the total cost of an ice cream cone, a hot dog, four hamburgers, a milk shake, and two deluxe cheeseburgers?

Page 43

hot dog = $1.00	cola = $1.00
order of French-fries = $1.00	ice cream cone = $1.50
hamburger = $2.00	milk shake = $2.00
deluxe cheeseburger = $3.00	taco = $2.00

1. $10.50 — Billy wants to buy three colas, three milk shakes, and an ice cream cone. How much money will he need?

2. $5.50 — If Sandra buys an ice cream cone, five deluxe cheeseburgers, three colas, two orders of French-fries, and three hot dogs, what will her change be if she pays $30.00?

3. $5.00 — What is the total cost of two hamburgers and an order of French-fries?

4. $13.00 — Jackie purchases three colas, two hot dogs, and four deluxe cheeseburgers. If she had $30.00, how much money will she have left?

5. $34.50 — What is the total cost of an ice cream cone, five colas, four milk shakes, five hamburgers, and five tacos?

6. $10.00 — What is the total cost of five milk shakes?

7. $4.80 — What is the total cost of three hamburgers if there is a 20 percent discount?

Page 44

hot dog = $1.00	cola = $1.00
order of French-fries = $1.50	ice cream cone = $1.50
hamburger = $2.00	milk shake = $3.00
deluxe cheeseburger = $3.00	taco = $2.00

1. $4.00 — What is the total cost of five colas if there is a 20 percent discount?

2. $15.00 — Marcie wants to buy three hot dogs, two tacos, and four hamburgers. How much will it cost her?

3. $10.00 — Jackie purchases four hamburgers and two hot dogs. How much money will she get back if she pays $20.00?

4. $3.00 — What is the total cost of three colas?

5. $10.50 — If Brian buys five hamburgers, two milk shakes, two deluxe cheeseburgers, and five orders of French-fries, how much change will he get back from $40.00?

6. $20.80 — What is the total cost of four tacos, a hamburger, three hot dogs, four deluxe cheeseburgers, and a cola if the items are on sale for 20 percent off the regular price?

7. $5.50 — If Allan buys a hot dog, five orders of French-fries, and four ice cream cones, and if he had $20.00, how much money will he have left?

hot dog = $1.00	cola = $1.00
order of French-fries = $0.50	ice cream cone = $1.50
hamburger = $2.50	milk shake = $3.00
deluxe cheeseburger = $3.00	taco = $2.00

1. $3.00 — If Jennifer buys four tacos and three deluxe cheeseburgers, how much change will she get back from $20.00?

2. $4.00 — What is the total cost of five colas if there is a 20 percent discount?

3. $20.00 — What is the total cost of a cola, five hamburgers, three deluxe cheeseburgers, and five orders of French-fries if the items are on sale for 20 percent off the regular price?

4. $26.50 — What is the total cost of three colas, three milk shakes, two deluxe cheeseburgers, three tacos, and a hamburger?

5. $5.50 — Michele purchases three ice cream cones. How much change will she get back from $10.00?

6. $10.50 — Jake wants to buy two milk shakes, three hot dogs, and three orders of French-fries. How much will he have to pay?

7. $11.00 — What is the total cost of a cola and five tacos?

hot dog = $1.50	cola = $1.00
order of French-fries = $0.50	ice cream cone = $1.00
hamburger = $2.00	milk shake = $2.00
deluxe cheeseburger = $3.50	taco = $2.50

1. $13.00 — If Audrey buys three colas and four ice cream cones, what will her change be if she pays $20.00?

2. $4.00 — If Paul buys four orders of French-fries, five colas, and two milk shakes, what will his change be if he pays $15.00?

3. $7.00 — If Adam buys five tacos, five colas, four orders of French-fries, three milk shakes, and five hot dogs, how much change will he get back from $40.00?

4. $2.50 — Sandra wants to buy a hot dog and a cola. How much will she have to pay?

5. $17.50 — If Janet wanted to buy three milk shakes, four hamburgers, an order of French-fries, and two hot dogs, how much would it cost her?

6. $20.50 — What is the total cost of three deluxe cheeseburgers and five milk shakes?

7. $30.00 — What is the total cost of four orders of French-fries, four milk shakes, five colas, four tacos, and five ice cream cones?

hot dog = $1.50	cola = $1.00
order of French-fries = $0.50	ice cream cone = $1.50
hamburger = $2.00	milk shake = $2.50
deluxe cheeseburger = $3.00	taco = $2.50

1. $5.50 — If Sharon buys two milk shakes, three tacos, two deluxe cheeseburgers, and four hot dogs, how much change will she get back from $30.00?

2. $2.00 — If Adam wanted to buy four orders of French-fries, how much would it cost him?

3. $3.00 — If Brian wanted to buy three colas, how much would it cost him?

4. $8.00 — What is the total cost of four hamburgers?

5. $10.00 — Ellen wants to buy an ice cream cone, a cola, three hot dogs, and a deluxe cheeseburger. How much will it cost her?

6. $9.00 — If Steven buys two orders of French-fries, three hamburgers, four deluxe cheeseburgers, and two colas, what will his change be if he pays $30.00?

7. $9.50 — If Janet wanted to buy four orders of French-fries and three milk shakes, how much would she have to pay?

hot dog = $1.50	cola = $1.00
order of French-fries = $0.50	ice cream cone = $1.50
hamburger = $2.00	milk shake = $2.50
deluxe cheeseburger = $4.00	taco = $2.00

1. $11.50 — If Allan buys two hot dogs, five orders of French-fries, four tacos, three milk shakes, and five ice cream cones, what will his change be if he pays $40.00?

2. $14.00 — Jackie purchases three hamburgers. If she had $20.00, how much money will she have left?

3. $11.50 — Paul purchases three ice cream cones, two deluxe cheeseburgers, and four hot dogs. How much money will he get back if he pays $30.00?

4. $7.00 — What is the total cost of three tacos and a cola?

5. $7.00 — Ellen wants to buy two orders of French-fries, a taco, a cola, and two ice cream cones. How much will she have to pay?

6. $27.00 — Michele wants to buy an order of French-fries, four hot dogs, four deluxe cheeseburgers, three colas, and an ice cream cone. How much will it cost her?

7. $20.00 — What is the total cost of five deluxe cheeseburgers?

hot dog = $1.00	cola = $1.00
order of French-fries = $0.50	ice cream cone = $1.50
hamburger = $2.50	milk shake = $2.00
deluxe cheeseburger = $3.50	taco = $2.00

1. $7.00   If Steven buys three hot dogs, and if he had $10.00, how much money will he have left?

2. $18.50   Allan wants to buy an order of French-fries, two colas, four hamburgers, and three tacos. How much will it cost him?

3. $4.00   Marcie purchases five tacos and a cola. If she had $15.00, how much money will she have left?

4. $5.50   Billy purchases two hamburgers, a deluxe cheeseburger, and four ice cream cones. How much change will he get back from $20.00?

5. $23.50   What is the total cost of a hot dog, five milk shakes, two colas, four tacos, and a hamburger?

6. $10.50   If Audrey wanted to buy five orders of French-fries and four tacos, how much money would she need?

7. $22.50   If Michele wanted to buy five hamburgers and five tacos, how much money would she need?

hot dog = $1.50	cola = $1.00
order of French-fries = $1.00	ice cream cone = $1.00
hamburger = $2.00	milk shake = $2.00
deluxe cheeseburger = $3.50	taco = $2.00

1. $16.00   Adam wants to buy two ice cream cones, two hamburgers, and five tacos. How much will it cost him?

2. $7.00   Allan purchases three orders of French-fries. How much money will he get back if he pays $10.00?

3. $17.50   What is the total cost of five deluxe cheeseburgers?

4. $12.50   What is the total cost of three deluxe cheeseburgers and two orders of French-fries?

5. $27.00   Paul wants to buy three hamburgers, a taco, five ice cream cones, and four deluxe cheeseburgers. How much will he have to pay?

6. $4.00   What is the total cost of two milk shakes?

7. $20.00   If Billy wanted to buy an ice cream cone, four milk shakes, three hamburgers, and five colas, how much would it cost him?

Made in the USA
Monee, IL
13 August 2025

23181984R10044